iPHONE 14 USER GUIDE

The Most Comprehensive and Easy-to-use and All-in-one Manual to Use your New iPhone with Tips and Tricks for Beginner and Senior Users.

Randall Klever

© Copyright 2022 by **RANDALL KLEVER** - All rights reserved.

This book makes an effort to offer trustworthy and factual information regarding the issue and situation at hand. The publisher agrees not to be obligated to provide an accounting, official authorization, or other qualifying services in exchange for the book. Assume you require legal or other expert guidance. The American Bar Association Council and the Committee of Magazines and Associations must both have authorized a Statement of Principles in order for a practicing professional to be allowed to do so. This content may not be copied, duplicated, or transferred in any form, whether digital or printed.

No part of this release may be reproduced, and storage of this information is only permitted with the author's prior consent. All intellectual property rights remain the owner's. Because the receiving reader is fully and completely responsible for any liability resulting from the use or abuse of any policies, procedures, or instructions included herein, whether as a result of carelessness or for any other reason, the information provided is presumed to be true and reliable. The publisher disclaims all liability for any compensation, losses, or damages coming directly or indirectly from the use of the material included herein. Any copyrights that the publisher does not hold exclusively belong to the writers.

This information is generic and only meant to be used for educational purposes. The information is given without the security of a contract or other assurance provision. The trademarks are utilized without the trademark owner's authorization or support, and the mark is advertised without their approval. The brands and logos mentioned in this eBook are the property of the respective owners and are in no way connected to them.

Table of Contents

Table of Contents	3
Chapter 1: - Top Features of the iPhone 14	6
Chapter 2: - Terminology	19
Chapter 3: - iPhone 14 4 Models Lineup	25
Chapter 4: - Set Up Your iPhone	28
Chapter 7: - Setting Up Apple Pay	60
Chapter 8: - Apps	67
Chapter 9: - Camera Basics	75
Chapter 10: - Siri	83
Chapter 11: - First-Hand Tips and Tricks	89
Final Words	104

Introduction

Have you ever wondered what goes on in a brand-new iPhone's manufacturing process? If so, you should read this user manual! With a number of upgrades and new features, the new iPhone 14 is more user-friendly than ever. But before we can truly understand what's new, we need to understand the parts that make up the iPhone and how they function as a whole. We'll go over each component of the iPhone 14 in depth and discuss how each one contributes to a wonderful user experience.

Your iPhone is one of the most important tools you can own. You know how to use it. But the thing is, you don't always know how to use it perfectly. The new iPhone is a great device, but you're going to want to know how to get the most out of your new phone. So whether you're upgrading to a new iPhone or you're thinking about getting a new iPhone, this guide will show you exactly how to get started with your new iPhone.

This user guide is all about what's new on iPhone, and how to make the best use of your iPhone device. It includes a thorough rundown of everything that's changed since iPhone. The iPhone 14 is here! There are lots of cool new features on this phone, like Face ID, Animoji, Portrait Mode, and more. This guide will help you use them.

What has the iPhone 14 added? A few things include a headphone jack, Face ID, better battery life, the brand-new wireless charging case, and a more streamlined appearance. Three key modifications are present in the new iPhone: a better display, a new body that is waterproof, Apple included a more potent camera system, as well as an enhanced processor and an additional speaker. These changes are all good news.

Everything you need to know about the new iPhone, including new features, is covered in the iPhone 14 user manual. What role will it play in your life? How does it feel to use?

Look no further if you're looking for a quick guide to the new iPhone 14. The iPhone 13 has previously been discussed, and now the newest iPhone has arrived. We'll discuss the new design, the best apps to download, what to anticipate from Siri, and how to get the most of your new iPhone in this book.

Chapter 1: - Top Features of the iPhone 14

Extra Storage

The good news is that Apple now includes 64GB as standard, which is excellent if you're always testing storage capacity. This is the same as doubling the internal memory of the iPhone 14 Additionally, Apple has ceased production of the 128GB model in favor of the 256GB iPhone 14.

It's also crucial to remember that iOS 14, which will be ready in a few months, will automatically reduce the size of photographs taken with the iPhone's camera, providing users a ton of more space.

Wireless Charging

Apple's newest smartphone has WiFi charging built in, so all you need to do to refuel it is set it down on a suitable pad. The iPhone 14 leverages the well-established Qi environment, thus it can be used with the majority of accessories available.

New Colors and Cup Design

Apple offers the iPhone 14 in a number of hues, including silver, gold, and space gray. The iPhone 14 and iPhone 14 Plus also feature a gorgeous cup-back design, which Apple claims is the strongest cup ever found in a smartphone.

A deep depth of color with exact hue and opacity is provided via a 7-layer color process and a color-matched aerospace-grade aluminum bezel. Additionally, the iPhone 14 and iPhone 14 Plus will be water- and dust-resistant.

A11 Bionic Processor

The newest brand push by Apple for the iPhone might be the fastest one ever. The A11 Bionic chip performs graphics around 30% faster than the previous brain found in the iPhone 14 on average. If true, it has a much larger likelihood than its predecessor of outperforming all other Android competitors in 2017 and the first part of 2018.

Interesting iPhone 14 Features

Powerful Picture Editing

The iPhone 14 has a better camera in addition to Apple's new A11 Bionic CPU, which is extremely powerful. These features make the iPhone 14 a fantastic tool, even for seasoned photographers. Numerous photo-editing programs are available on the App Store. With the right equipment, you can take gorgeous photos that seem professionally done. Photofox is one of my favorite software programs. The finest features of desktop programs like Adobe Photoshop are merged with the ease of editing and enhancing on a smartphone.

The ability of the program to edit photographs in levels may be quite useful for professional photo editing. Using visual components, innovative designs may be made relatively quickly. Although it is often free, the application form offers in-app purchases.

Portrait Lighting for Portraits

With the release of the iPhone 14 Plus, Apple debuted the incredible Face photo setting. Face Lighting, however, requires an upgrade for the iPhone 14 Plus.

In a nutshell, the Face shot option lets you to take images of a subject that are extremely crisp against a background that is blurred. The iPhone 14 Plus, on the other hand, enables you to increase the number of lights to resemble a real studio. The best feature is that after taking a Portrait mode photo, you can still change the lighting so that it exactly matches the one you want to purchase.

By selecting Edit in the impression in the Photos app, you can adjust the lighting in the already-taken portrait photo. When it happens, you can adjust the image settings by swiping the Face Lighting wheel.

Screen Capture

The Control Center now offers a new feature called native screen recording that is built straight into the most recent iPhone models. You may access the Control Center by swiping up from within the screen. Third-party apps are no longer required thanks to this capability. Unfortunately, you must go to the iPhone's configurations/settings page in order to access the display documenting icon; it is not by default visible in the Control Center.

The iPhone 14 and iPhone 14 Mini

Apple Inc. produces, designs, and markets the iPhone 14 and iPhone 14 Mini as two different smartphone models. They are the more affordable 15° generation iPhones, which were created as the iPhone 12's replacement. They were revealed alongside the pricey iPhone 14 Pro and iPhone 14 Pro Max flagships at a virtually orchestrated Apple Special Event on September 14, 2021, at Apple Park in Cupertino, California. On September 17, 2021, Apple opened pre-orders for the iPhone 14 and iPhone 14 small smartphones; both devices went on sale on September 24, 2021.

At first glance, the iPhone 14 smartphone doesn't appear to be all that unique. Many people might mistake it for a minor update due to its uncanny resemblance to the iPhone 12's design and flat frame, which is typical of Apple's most recent models. But that's not the case, as we'll see in a moment.

It varies from its predecessors in a few ways, most notably by having buttons that are placed much lower on the device and by having Face ID and a True Depth sensor. Even though most people have a weirdly contradictory relationship with this feature, it also has larger back cameras that are placed diagonally and a notch that is 20 times smaller. However, when you use this cutting-edge smartphone and take in its gorgeous display, which is brighter than any of its predecessors, you'll find that it provides much more than simply a smaller but larger notch.

The battery life has increased. Another important improvement made by Apple to an already superb camera is a function called as computational photography, which gives you a sense of professionalism when using the camera. Additionally, there is a larger storage capacity in addition to the world's fastest chip, the A15 bionic chip. You did indeed read it right. Now let's talk about the details.

iPhone 14 Mini

I'm sorry to disappoint you, but if you believed that pocket-sized phones were obsolete, I have news for you: the iPhone 14 Mini is leading the charge. The iPhone 14 Mini is fantastic on its own, regardless of size. It has even been considered as by some as the most potent tiny phone ever. That's a bold claim to make for a phone, but perhaps you'll understand why later.

Apple has saved a few noteworthy features for the iPhone 14 Pro and iPhone 14 Pro Max, but the iPhone 14 Mini gets a ton of nice updates. A15 Bionic processing speed, a brighter display, and intriguing new camera functions like video's Cinematic mode are among them.

Unboxing iPhone 14, iPhone 14 Mini, iPhone 14 Pro, and iPhone 14 Pro Max

For all iPhone 14s, the unboxing procedure is essentially the same. Accordingly, the iPhone 14, iPhone 14 mini, iPhone 14 Pro, and iPhone 14 Pro Max are all packaged in light boxes.

The lack of a power brick in the package is the reason of this. The plastic shrink-wrap from the iPhone 14s was also removed by Apple, which the company claims has prevented the use of up to 600 metric tons of plastic.

To achieve zero waste, Apple currently makes use of recycled cardboard and aluminum in many of its packaging materials. The included accessories have not seen a significant amount of change. An instruction handbook, a SIM card tool, a white Apple label, and a lightning to USB-C cable are included.

Set Up and Get Started

The Apple logo is displayed in the screen's center on a dark background as soon as the iPhone is turned on. After that, a multilingual "hello, welcome" background will appear to greet you.

The following choice is "Select a Siri voice." "Hello, I'm Siri," will be the first thing you hear. Select the voice you prefer. Just select one of the four alternatives that are provided. Naturally, you can always modify this in the settings afterwards.

Switch iPhone Off

- iPhone with Face ID: Press the side button and hold each volume button simultaneously until the display of the slider, and then drag the Power Off slider down.

- Home button iPhone: Press and hold a side button (depends on your pattern) and drag the slider. Sleep/Wake button.

- Illustrations of 3 iPhone models, all facing screens: On the left side of the gadget, the picture depicts the up and down volume buttons. The side button on the right is displayed. The center picture shows the side button on the right side of the gadget. The graphic on the top of the smartphone depicts the Sleep/Wake button.

- All models: Enter Settings > click General > press shut down, then drag the slider.

- Switch on iPhone.

- Touch and hold the Sleep/Wake button or side button until the Apple logo is shown (depending on your model).
- Force restart.
- Do the following:
    - Press the volume up button, and release it fast.
    - Next, press and hold down the side button. Release the button if the Apple logo is displayed.

An iPhone device that lacks a home button is shown facing up. Instead, the device's left side displays the up and down volume controls, and there is also a second button visible there.

Automatically Update iPhone

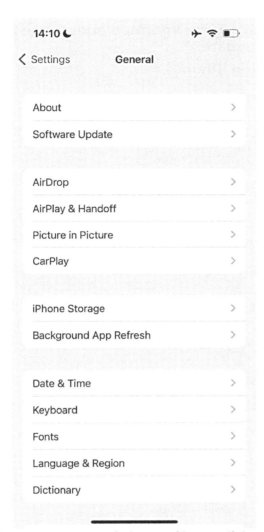

When you first set up your iPhone, if you did not enable automatic updates, perform the following:

- Go to Settings, then pick General, then Software Update, then choose Automatic Updates.
- Download and install iOS updates after turning on iOS updates.
- If an update is available, the iPhone will download and install it throughout the course

of an overnight WiFi connection and charging period. You are informed prior to the installation of an update.

Manually Updating iPhone

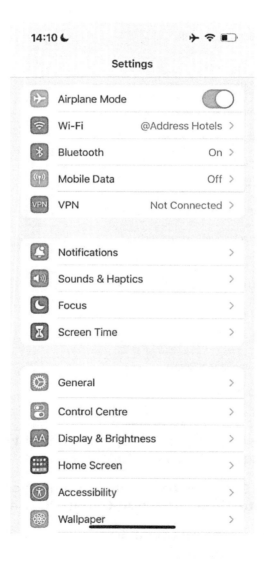

You can always check for the software updates and install them.

- Update software by going to Settings > General.
- The monitor displays the version of iOS that is currently installed as well as whether an update is available.
- To disable automatic updates, go to Settings > General > Tap Software Updates > Automatic Updates.
- using iCloud as a backup
- Backup for iCloud can be found under Settings > [your name] > iCloud.
- Turn on the Backup iCloud feature.
- Whenever your iPhone is turned on, locked, and connected to Wi-Fi, iCloud backs it up automatically each day.

Note: On 5G-capable devices, your carrier may provide you the option to backup your iPhone using a cellular network. Select Settings > [your preferred name] > Turn backup on or off under iCloud > iCloud Backup.

- For a manual backup to start, tap Back Up Now.
- Go to Store management > Backups under Settings > [your name] > iCloud. Choose a backup and then select Backup delete to eliminate it from the list.

Note: When an iCloud function is enabled (like iCloud photos and contacts) under settings > [your name] >

iCloud, Its information will be kept on iCloud. It is not included in your iCloud backup because all of your information is constantly current on all of your devices.

Back-Up with Mac

- Connect your wire to your PC and iPhone simultaneously.
- Pick your iPhone from the Finder sidebar on your Mac.
- On MacOS 10.15 or later, the Finder must be used to backup the iPhone. Earlier versions of macOS and iTunes can be used to backup the iPhone.
- The top menu of the Finder window will now read "General".
- Select "Back up all iPhone data to Mac" only once.
- To encrypt and password-protect backup data, choose "Encrypt local backup."
- the Back Up button.

Note: If you're running Wi-Fi, you can also wirelessly connect iPhone to your computer.

- iCloud backup for restoring an iPhone
- Turn on a brand-new or recently wiped iPhone.
- To choose a language and location online, adhere to the directions.
- Click Manual Set Up.
- Tap on iCloud Backup restore and adhere to the instructions displayed.
- iPhone backup restoration
- Connect your backup PC via USB to your recently erased iPhone.

Do one of these:

- In the Finder sidebar on your Mac: After selecting your iPhone, click Trust.

In order to restore the iPhone from the backup using the Finder, MacOS 10.15 or later is required. To restore from a backup with earlier macOS versions, use iTunes.

- If you have multiple connected devices to your Windows PC and are using the iTunes application, click the icon in the top-left corner of the iTunes window to choose the recently destroyed iPhone from the list.

- On the welcome screen, click the "Restore from this backup" option. From the list that appears, select your backup, and then click Continue.

- If your backup is encrypted, you must provide the password in order to restore data and settings.

Send Items with Airdrop

- Open the item, then click the Share icon or another button that displays the app's sharing options.

- Press the AirDrop icon in the share options row, then touch the user's profile picture.

Tip: For the iPhone 11 and iPhone 12 models, point your iPhone in the direction of a different model and click the user's profile photograph at the top of the screen.

- If the recipient does not appear to be an AirDrop user, ask them to open the Control Center on their iPhone, iPad, or iPod touch and turn on the AirDrop receiving feature. To send someone on a Mac, ask them to set up AirDrop so that they are discoverable in the Finder.

- enabling AirDrop on your iPhone so that others can send you stuff

- Open the Control Center, tap and hold the set of controls in the top-left corner, and tap the AirDrop icon.

- Tap Contacts only or everyone to choose who should send you goods.

- As requests come in, they may be approved or rejected.

Chapter 2: - Terminology

Performance and speed, which have emerged as the clear leaders of the industry, have surpassed other cellular technology features from the perspective of the consumer. Consumers today are more involved in the operation of the mobile devices that all companies introduce onto the market each year than users of prior generations. Why? Since the invention of gadgets with touch screens, there has been a kind of design universality. Today, it is challenging to find a mobile device that differs considerably from the competition. In contrast to what happened with prior technology generations, the external design is now less important (where an inventive presentation and a radical format were the attractiveness that corporations offered). Regardless of the equipment's design, today's users look for performance and quickness.

This chapter contains all the information you need to know about the new iPhone 14. That is, its most important and vital technical aspects so that you can make the best decision with more understanding. Going forward, you should pay extra attention to what you read on these pages because we'll go into great detail on the upcoming iPhone 14 model from Apple.

Technical Specifications

We want to answer any queries you may have about Apple's latest suggestion. We shall therefore start with the product's technical details.
- 146.7 X 71.5 7.7 mm are the measurements of the apparatus. 174 gram weight.
- Body of the equipment: An aluminum frame with a glass front screen and glass back panel.
- Blue, Pink, Midnight, Starlight, and Red are colors that are readily accessible on the market.
- a 3240 mAh Li-Ion equipment battery (non- removable

battery). With multimedia playback, the battery life is predicted to last up to 19 hours and with music playback, up to 75 hours.
- With Super Retina XDR OLED technology, regarding the Screen
- Touch.
- 16-bit color depth
- 90.2 square centimeters is the size of the screen.
- 6.1 inches overall in length.
- Ceramic glass with a scratch-resistant oleophobic cover
- Concerning the Camera
- a rear camera with 12 megapixels (double camera).
- Dual-pixel PDAF, sensor-shifts OIS, and 12-megapixel wide-angle option.
- a 12-megapixel wide-angle option.
- additional GSM, CDMA, HSPA, EVDO, LTE, and 5G network.
- 5.0 Bluetooth
- GPRS.\sEdge.
- USB.\sGPS.
- WiFi (hotspot, dual-band, 802.11 a/b/g/n/ac), 802.11 b/g/n.
- no headphones (headset connection via Bluetooth).
- Face ID sensors for identification (Face Recognition, Accelerometer, Gyroscope, Proximity Sensor, Compass, and Barometer).
- charging via USB 23 W for a 2.0 quick charge. Thirty minutes to a 50% increase.
- mobile charging
- Apple Pay.

Design

The first thing you should be aware of is that the design of the iPhone 14 is rather similar to that of previous models. One of the first things we notice about it is how narrow its stripe is. The top is higher despite this, though. This function is quite useful since it allows you to present information more precisely. On the other hand, the device is slightly thicker and heavier than previous incarnations, although scarcely being noticeable in everyday use. Its thickness is not a disadvantage as compared to prior models because it is invisible during normal use.

As we can see, the cameras on the rear panel are now placed diagonally rather than vertically. Similar to this, the iPhone 14 lacks all of the technical markings that were present on earlier iPhone versions. The differences offered by this edition, although adapting to the demands of its customers, are hardly distinguishable from those of other editions. In this aspect, there is absolutely nothing to criticize, and Apple is active in adapting its products to the most recent trends.

Let's Talk About the Screen

The 6.1-inch Super Retina XDR OLED screen on the iPhone 14 stands out above products from other notable companies on the market and is not at all a flaw. Also deserving of special attention is its 2532 x 1130 pixel resolution. For those who prefer it, the iPhone 14 features a brighter display than the model that was introduced to the market in 2020. The fact that this model now achieves 13,800 nits is significant and must not be disregarded. The ability to utilize technology safely outside in the sun is highly valued by users today. With an iPhone 14 in your palm, you can do it with ease.

The screen has Ceramic Shield protocol protective glass on it for prevention. We were able to see during the initial testing that this method makes the iPhone 14 significantly more scratch-resistant than any other equipment, even the best technology of other manufacturers. Every user, but notably the more reckless ones, values this perk highly.

Cameras

Everyone is aware that iPhone cameras are typically in every way superior to those on devices from other manufacturers. In actuality, this benefit is the one that most appeals to the younger user demographics that favor creating content and using social media. Of course, the photographic capabilities of the iPhone 14 are superior to those of older versions. With the addition of matrix stabilization, the camera can now collect about 45 percent more light. You will certainly be able to capture stunning shots with this iPhone thanks to its technological features and extra features, like the range of modes (portraits, night photos, landscapes, among others).

Another movie that has gotten good reviews is The Cinema Effect. What is the issue? When making a video, you might begin by concentrating on something close-by before moving on to something farther away. Use only one touch on the screen. It varies from past models in that it is capable of recording video in 4K resolution at a 60 frame per second. You won't be surprised by these enhancements if you routinely use the brand. If this is your first iPhone, however, the photography section will provide you with all the guarantees of happiness.

Performance

Performance is the most important factor for modern technology consumers. With so many options, it's easy to become disoriented and make decisions that are counterproductive to our objectives. With Apple, performance is always guaranteed. With the exception of the first failures a few generations ago, the iPhone has been significantly improving in this area. Performance of the iPhone 14 in this instance is dependent on its internal A15 Bionic CPU. It is undeniably a key advantage that this phone operates quite rapidly and agilely, as we have confirmed. You may play games, edit videos, and do difficult tasks with the help of specialized software like iMovie. However, Apple's mobile processors have always been a cornerstone of its innovations. The findings and artificial tests that we have put the new model through demonstrate to us remarkable processing capacity, just what the contemporary user need to handle all the obligations, needs, and chores of daily life.

If you thoroughly consider the functionality of the equipment before making a decision, you won't be dissatisfied. Apple and its iPhone model continue to do well in the most specialized reviews when it comes to the speed and quickness with which the CPU supports functions in harmony, regardless of their complexity. Unquestionably a fantastic opportunity in these times. Yes, you can still start using an iPhone.

Battery Life

Some businesses are forced to choose between speed and load capacity as a result of the advancement of quicker, more flexible CPUs. With the iPhone 14, this does not take place. This device will provide you with more battery life when fully charged, which indicates a major advance over the model that was introduced to the market in the year 2020. That is to say, the device now has a 2 hour longer working time, an accomplishment that deserves to be praised and that is cited as a crucial factor in the reviews of the most qualified experts in this field. A non-removable lithium-ion battery powers the gadget. Its battery capacity is 2340 mAh. Results from the tests have been overwhelmingly good, particularly when using the 20 W quick charging technology. This means that in just over 30 minutes, it will be possible to restore up to 50% of the device's charge.

Chapter 3: - iPhone 14 4 Models Lineup

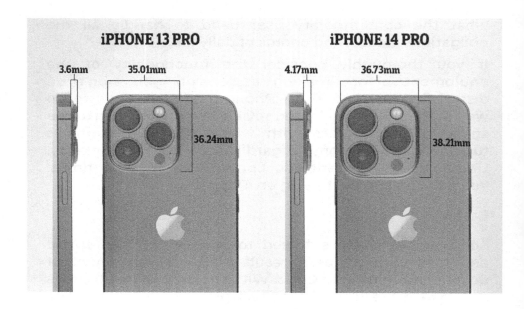

Overview of the iPhone 14 Mini

The 13 Mini is the smallest and least expensive member of the iPhone 14 series. The size is the only variation in the features between the iPhone 14 Mini and the iPhone 14. The iPhone 14 Mini is the best option for anyone looking for a compact, lightweight, and portable flagship iPhone.

The battery of the iPhone 14 Mini has a notable drawback as well because of its little size. As a result, it might not endure as long as the physically larger variants. Additionally, typing on the relatively small 5.4-inch display is more challenging than on its bigger predecessors.

Overview of the Standard iPhone 14

The most popular gadget is the iPhone 14, which is neither too small nor too large. It uses OLED technology in its 6.1-inch, 60Hz display, just like earlier models.

Overview of the iPhone 14 Pro

The 6.1" screen size of the iPhone 14 Pro is carried over from the iPhone 12 Pro. But more importantly, it has a dynamic frame rate display that can exceed 120Hz depending on the data being displayed. It also has larger internal storage. It can automatically transition between a low refresh rate to prolong battery life and its maximum refresh rate of 120Hz because it is dynamic. Its user interface (UI) elements and contents that support 120Hz (like games) appear to move more fluidly as a result.

Overview of the iPhone 14 Pro Max

The iPhone 14 Pro Max, as suggested by its name, is the biggest and priciest model in the 2021 iPhone lineup. Compared to the 6.1-inch screens of the 13 and 13 Pro, the Pro Max's screen is 6.7 inches large. The display also has a higher resolution and a 120Hz variable refresh rate.

Additionally, the iPhone 14 Pro models offer the most possible storage capacity in comparison to other iPhone models made available to date (1TB). The 13 Pro Max with 1TB of storage is also the most expensive iPhone to date, it should be emphasized.

Should you go from an Android handset to an iPhone 14?

Any Android phone will first feel somewhat different from an iPhone. Because Android is a more flexible operating system that allows users to download apps from sources other than the Play Store, Google's version of the App Store, there is a difference in the user experience and feel between the two operating systems.

Apple claims that iOS 15, which comes preinstalled on the iPhone 14 series, is a particularly secure and privacy-protecting software. iPhone users, in comparison, may only download pre-approved apps from Apple's App Store and have significantly less customization possibilities.

Although this is limiting, it is also an important security feature because it makes it more harder for an iPhone user to mistakenly download and install malware. Apple's "App Monitoring Transparency" feature in iOS 14 also gives iPhone owners the option to turn off tracking of their devices.

In comparison to Android phones, an iPhone may have many more years of software updates. This implies that you might only receive two years' worth of enhancements, depending on the manufacturer of your Android phone.

Because Apple has a reputation of providing software support for more than four years, the new iPhone is without a doubt the best choice if you want to buy a phone to use for a long time with the most recent operating system.

Chapter 4: - Set Up Your iPhone

Setting up your iPhone is the first thing you need to do when it initially turns on. When you first open the smartphone's box and turn it on, you'll see it start to display "Hello" in a number of different languages.

Then, after turning it on, choose your favourite language from a list of choices.

You will next be asked to choose your residence's nation or region.

Quick Start

The following setup screen is the "Quick Start" setup page. In order to access the installed apps and data on your previous iPhone, you may either manually configure the device or sync and link it to your new one.

Once you've finished this step, a screen will appear requesting you to wait for your iPhone to activate for a few minutes.

Configuring Face ID and Privacy Settings

The following screen is for data and privacy settings. You can choose the privacy settings you want to enable here to give your device additional security.

When you have finished with the Data & Privacy page, the Face ID setup screen will appear.

You can choose to configure it now or later by choosing Set up Later from the Settings menu. Clicking Continue will start scanning your face and saving the data so that you may use it to unlock your iPhone later on if you wish to set up Face ID straight away. Try to position your face inside the circle when the face-scanning procedure begins. After then, start turning your face toward the spinning motion.

Setup of Apps & Data

The Apps & Data tab will show up after Face ID setup is complete. How you transfer your current iPhone 14's apps and data is up to you. To do this, you may use a number of different techniques.

iCloud Backup Restore

Your data can initially be restored from the most recent iCloud backup. The final option is to immediately transfer all of the pertinent information from your old iPhone. An alternative approach is to transfer all relevant data from your old Android smartphone to your new iPhone. You can also decide to forego this data transfer entirely.

Express Settings

Once you've chosen whether to transfer current data or move on to the next step, the Express Settings screen will display. Here, features include Siri, Maps, and the right to gather user data for analytical purposes.

Once you've selected your selections, you may touch "Continue" to move on to the following screen or modify them.

Inserting Your SIM Card

You must first place your SIM card into your new phone before turning it on for the first time. This is necessary since the software won't recognize your network provider and establish a correct connection if your device doesn't have a functional SIM inserted. Additionally, if your Apple ID account has been compromised, the SIM card installation process will stop any potential data loss.

To do so, follow these steps:

First, ensure that you're using the right size SIM card from your manufacturer (Apple recommends only using micro-SIMs). When ordering a SIM at the store, make sure they give you one that fits perfectly inside the slot. You should be able to see the notch where the SIM goes when looking down at your iPhone 14. Do NOT force a larger-sized SIM into the slot! Doing so could damage your phone.

Next, remove your SIM tray by sliding off its metal tab (with a SIM ejector) located beside the display assembly. Slide your SIM card into the slot. It should fit snugly within. Make sure the SIM is fully seated.

Turn on your iPhone again and wait for the initial boot-up sequence to complete. Afterward, you should now be connected to your cellular carrier. From here, you can start making calls, sending texts, surfing the web, etc...

Home Screen Setup

Home screens have been decried for years as a necessary evil. You can select from a variety of home screen designs and layouts, and some of them might even come with extra "bells and whistles" for your home screen.

It's crucial to know how to customize your iPhone home screen to help you get the most out of your smartphone, even though you might not always want to go to the work of rearranging the icons.

Personalizing Your iPhone 14 Home Screen

As you personalize your iPhone home screen, you can create a custom set of icons that can easily be rearranged or deleted anytime you want.

When you open the app you want to use, it should be the top app on the home screen. Click Edit on the screen that pops up.

Click "Add icon." You'll be prompted to pick a photo for your new icon, just like you'd do for an album in your photo library. There's also a checkbox to add a note next to the icon.

The icons along the top are apps and the two icons at the bottom are your home screens. You can drag an app or home screen down to customize the order in which they appear. Click Done to confirm the new order.

How to Use Apps and Home Screens

Let's now examine how to utilize the home screens and apps. Just press an app's icon to start it. Your list of home screens will appear when you swipe up from the bottom of the screen; tap the one you want to use.

When an app is open, you can utilize it as you normally would. Home screens can be added or taken away, and you can scroll between them to access another.

Additionally, you can make unique folders to organize your apps. Tap and hold the app you wish to group, drag it to another app, let go when it drops to form a new folder, and then repeat for all desired apps.

Using Home Button on iPhone 14

With the help of the Assistive Touch feature, people with certain disabilities can still use their phones to navigate without actually touching the screen. Although Apple first proposed the concept for this function in iOS 9.3, it has been progressively expanding its support over time. What you must do if you have never utilized Assistive Touch:

1. Open Settings.
2. Tap Accessibility.
3. Scroll down until you see Assistive Touch.
4. Enable Assistive Touch.
5. Now try to open any app or navigate between apps by tapping icons instead of swiping them.
6. You will now notice two small dots at the bottom right corner of each icon when they are tapped; these indicate where you should swipe to open the app.
7. To close the app, tap the dot again.
8. When you want to go back to the previous page, just swipe left from the right edge of the screen.

Using Your iPhone 14 as Hotspot

Apple has a new system-level feature for iOS 15 that makes your iPhone into a wireless router for Wi-Fi connections. With this, you don't need to jailbreak the iPhone to turn it into a Wi-Fi hotspot. Here's how to turn your iPhone into a Wi-Fi hotspot:

1. Open the Settings app on your iPhone.
2. Open the Personal Hotspot tab and click on Allow Others to Join.
3. Tap on the toggle switch note down the password.
4. The iPhone should now connect to your wireless network.

Using Gestures

Simply put, gestures give you more natural interaction with the device than tapping does. A built-in hardware button on the iPhone responds to a gesture by carrying out the same function as a tap. The biggest benefit of gestures is that they are relatively simple to use. They offer a very natural user interface, and the gadget can recognize the move you are making.

On the down side, gestures rarely allow the user to conduct numerous taps and are less accurate than touch taps. This discussion will go over various motions and how to use them.

How to Perform Gestures

On the iPhone, one of the two primary buttons can be used to make gestures. The home button is the only button, and pressing it brings up the device's home screen. The Sleep/Wake button is the additional button. The device either enters or exits sleep mode when you press it. There are a few techniques for using iPhone motions:

- Tap on the Home button. You can only make a gesture as simple as this. This move will also bring up the phone's home screen. But unless you are holding the iPhone, tapping the button does nothing to the hardware. For instance, if you are on a phone call, tapping the Home button won't wake the device but may end the call.

- Swipe upwards on the home button. This gesture, which requires that the device be unlocked, brings up the Phone app.

- Swipe downwards on the home button. This gesture, which requires that the device be unlocked, brings up the home screen.

- Swipe upwards and hold on to the home button. This gesture, which requires that the device be unlocked, brings up the Settings app.

- Swipe downwards and hold on to the home button. This gesture, which requires that the device be unlocked, brings up the Control Center.

Remember that using the right hand to perform gestures on the smartphone is the same as using the right hand to perform touch taps.

Using an external mouse and keyboard that simulates a trackpad, you may also use gestures on the iPhone.

The trackpad interprets your gesture as a tap when you use it.

Please take note that the iPhone does not interpret mouse motions. Utilize Apple's Magic Mouse if you want to use a mouse to make a gesture.

The iPhone's home button can be operated with a gesture even when it is not being held. If you need to conduct a long press on the home button or another motion, you can utilize the long-press control detailed below.

The Long Press

The long press is an extended version of a touch tap. This control does not require that the device be unlocked, but the gesture will not activate any hardware button if you are on a phone call.

To perform a long press on the home button:

1. The home button should be pressed with the index finger.

2. Lift your finger off the home button and let go of it after a brief moment.

This will be perceived by the iPhone as a prolonged press. It won't turn on right away; instead, it will do so after the device turns on for the next time as a touch tap.

This indicates that no additional gestures will be recognized if you push the home button while holding it down. When the iPhone is on a call, though, tapping the home button won't work.

To perform a long press while on a phone call:

1. Lift the phone's ring indicator off the ear and hold the home button down.

2. After a few seconds, lift the indicator of the ear and release the home button.

The phone will then ring or the home screen will be shown again. On the home button, you will be able to make more gestures.

Remember that if you are on a phone conversation, a lengthy press will only activate the home button and not any other physical buttons.

Long presses can also be used on the Sleep/Wake button. Then, depending on whether you are using a long press or a touch tap, the iPhone will either enter or exit sleep mode.

Activating the Night Shift Mode

You are probably aware of Apple's Night Shift feature for the iPhone, which automatically lowers the blue light generated by your device to improve the quality of your sleep. But how exactly do you do that? There are manual and automatic methods, though. I'll demonstrate how to activate or deactivate the Night Shift feature on your iPhone 14 in this part.

You must adhere to these easy instructions if you wish to enable Night Shift on your iPhone 14. Let's start now:

1. Open the Settings app.

2. Tap the General option at the bottom of the screen.

3. Now tap the Display & Brightness option.

4. Scroll down to the Night Shift option at the bottom of the screen.

5. Tap the Night Shift option at the bottom of the screen.

6. Also, tap the switch that will turn the Night Shift option off.

The settings are displayed here. On your iPhone, you may easily enable and cancel the Night Shift feature.

Screen Brightness Adjustment

You undoubtedly already know that the iPhone 14 has a built-in light sensor that is intended to automatically adjust the brightness of your screen to match the environment you are in. However, in most situations, you won't be able to view the screen under such circumstances, and you'll wind up needing to manually change the screen brightness. This isn't always a good thing, so we're going to walk you through how to change the brightness setting on your iPhone 14 on iOS 15 using the instructions below.

Your iPhone's screen will have a slider that you may use to change the brightness. It is situated close to the center of the screen, and you can change the brightness by sliding it to the right or left. You are currently using the maximum brightness level, which is the default option for your device, if the slider is visible on the right side of the screen. In the minimum brightness level, which is how the device will often be set, the slider should be visible on the left side of the screen.

On your iPhone, tap the Settings app to adjust the screen brightness. After that, you must press the General option that is visible to the right of the screen. This will bring up a new screen with a number of options, including Ambient light, Control brightness for accessories, and

Adjust screen brightness.

Several alternatives will appear when you tap on the option you want to change. The iPhone has five screen brightness settings that you may choose from: Normal, Dim, Night, Low, and Flash. You can access any of them by touching on the various options.

What Is True Tone? How to Use It

The first Apple iPhone had a black display with white letters on a black background. One of the first screens with a true-to-life color representation on a mobile device was the Retina display on the iPad, which Apple announced in 2013. This was caused by the display's sophisticated IPS technology, which matched the hue of outside lighting with the image on the screen.

Today's majority of Apple products feature a True Tone display that can adjust to the surrounding lighting conditions. This enables the gadget to faithfully reproduce text and images so that they appear to us as we would in the actual world.

What Is a True Tone Display?

The phrase "True Tone" refers to a display that modifies color temperature in response to ambient light, including natural light, artificial light, and even indoor and outdoor lighting. If the display isn't adjusted to the correct ambient light, the True Tone display uses this data to dynamically alter the display's color, producing a realistic-looking image.

You can find instructions on how to get True Tone for iOS devices in this section.

How to Get True Tone on iOS

When you open an app in iOS, there's a setting called Display & Brightness that can turn on/off True Tone. You can access this setting in Settings > Display & Brightness >

True Tone.

You'll see a toggle to turn True Tone on/off (this toggle will be grayed out if True Tone isn't on).

When this option is enabled, you'll see a small icon on the top right of the screen which indicates that you are viewing the display through a True Tone display.

Chapter 5: - Supported SIM for Your iPhone

When you utilize the compatible SIM(s) that are advised by the Carrier/Cellular Network Service Provider, inserting a SIM into your newly purchased iPhone is simple.

I stated the dual SIM that is compatible with your iPhone during the introduction portion. Both Nano-SIM and eSIM are used.

What Is Nano-SIM?

The smaller size SIM card, known as the NanoSIM, is substituted for the largest size SIM.

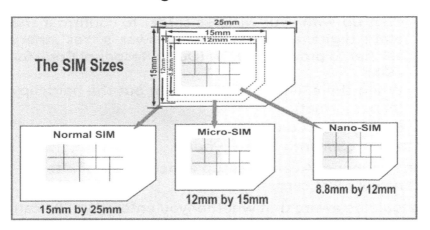

Nowadays, Cellular Service Providers (Carriers) are providing a 3-in-1 SIM. A SIM had been shaped into three sizes. Namely; Normal SIM is the Biggest size, Nano-SIM is the Smaller size, and Micro-SIM is the Smallest size.

It is only the Nano-SIM that can be used with your iPhone.

What Is eSIM?

The letter "e" stands for an electronic SIM, as suggested by the name, which may be digitally loaded on your iPhone by scanning the Quick Response (QR) for the eSIM Cellular Plan Setup using your camera.

You can therefore use your iPhone without a physical eSIM Card. Simply let your carrier know that you wish to add an additional eSIM to your iPhone, and they will offer one for you.

Soon, we'll go over how to set up the digital Cellular Plan on your iPhone.

How to Digitally Install eSIM on Your iPhone

- Chat up with your wireless carrier to confirm if she offers digitalized eSIM. If the answer is "Yes", then, ask her to provide you QR (Quick Respond) Code for eSIM.
- When the eSIM is provided, print out the hardcopy (paper format).
- Scan the QR Code.

How to Insert SIM into Latest iPhone

Your new iPhone 14 can be setup either before or after inserting your SIM card.

You should be aware that whether you enter your SIM card before setting up your iPhone or not, it will not stop your recently purchased iPhone from finishing the setup.

On the left side of your iPhone, in the center, is where the

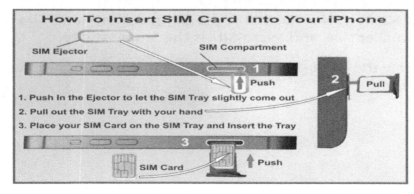

SIM compartment (place) is located.

How To Insert Your nano-SIM Into Your iPhone

- Let the screen of your iPhone face up and place it on a table.
- Insert the SIM Ejector into the hole on the SIM Tray on the left side of your iPhone after picking up the SIM Ejector.
- Apply some pressure to push the interior surface inward when the head of the ejector reaches it. The SIM tray will emerge somewhat right away.
- To fully remove the SIM Tray, use your two fingers.
- As seen in the image above, turn the nano-metal SIM's surface upward.
- The SIM should be turned so that it fits in the correct form on the SIM Tray before being placed on it. On the tray, make sure the SIM is well balanced.
- Pushing the Tray within the SIM slot will insert it.
- Check to see that the SIM Tray's surface is absolutely level (parallel) with the iPhone side surface. To verify the ideal level, move your finger across the tray's surface and across the iPhone's body.

How to Use Dual SIM on Latest iPhone

You can only choose one of the options listed below to use dual SIM on your iPhone.

1. On your iPhone, two working eSIMs are supported.
2. As an alternative, you can only utilize a

nano-SIM card and a working eSIM.

3. Two nano-SIM cards were expressly created in a region like China Mainland where eSIM is not offered. You may purchase the iPhone 14 in Macao and Hong Kong as well.

How to Digitally Install eSIM on Your iPhone
- Ask your wireless provider if she sells digitalized eSIMs to make sure. Ask her for the QR (Quick Respond) code for the eSIM if the response is "Yes."
- Print the hardcopy after receiving the eSIM (paper format).
- Detect the QR CodeAsk your wireless provider if she sells digitalized eSIMs to make sure. Ask her for the QR (Quick Respond) code for the eSIM if the response is "Yes."
- Print the hardcopy after receiving the eSIM (paper format).
- Detect the QR Code.

How to get a new eSIM through Wireless Carrier App

At the Homescreen Page: Tap on App Store Icon

On the Apple Store Page: Select your Carrier to download on your iPhone.

On the Carrier Page: Tap on the Carrier App Icon on your Homepage and buy a cellular plan for eSIM.

How to Add Your eSIM Cellular Plan Through Settings

Ask your wireless provider if she sells digitalized eSIMs to make sure. Ask her for the QR (Quick Respond) code for the eSIM if the response is "Yes."

Print the hardcopy after receiving the eSIM (paper format).

Detect the QR Code.

At the Homescreen: Tap on the Settings App

On the Settings Page: Tap on Cellular

On the Cellular Page: Tap on Add Cellular Plan

On the Add Cellular Plan Page: Place the Rear Camera on the eSIM QR Code to scan it, and continue the instruction promptly.

Another Method on

How can you activate the iPhone Camera app to scan QR codes?

In reality, you must enable scanning in your camera settings by doing the activation steps listed below before your iPhone will be able to scan the Quick Response.

How to Activate the Camera QR Code Scanner At the

Homescreen: Tap on the Settings icon.

On the Settings Page: Select Camera On

the Camera Page

- Tap on the Scan QR Code *activation switch* to activate it. The switch will change from white to green.

- Swipe up from the bottom center to go back to Homescreen.

How you can use the Camera App to Scan the QR Code on your iPhone

At the Lock Screen or Homescreen or Control Center: Tap on the Camera icon

On the Camera Page
- Verify that the Rear (Back) Camera is on. If not, tap the Camera rotating symbol (rotator) in the lower right corner of the screen to switch to the back camera.
- Place your camera so that the Quick Response Code is in the center.
- The Cellular Plan Notification will show up on the screen of your iPhone.
- Tap on the Notification.
- Tap on "Continue" at the screen bottom.
- Select Add Cellular Plan.

Note: When asked for a confirmation code to activate the eSIM, your carrier will issue you with a Code Number; enter it.

How to Enable Cellular Plan in the Control Center Your iPhone

At the Homescreen: Swipe down at the top right of the screen to launch the Control Center.

On the Control Center Page:
- Tap on the Cellular network activation icon. The upper left corner of the Control Center has two Cellular Status Icons stacked on top of one another.
- To get back to the Homescreen, tap the simple side or swipe up from the bottom.

Note: *The Carrier 1 will be using LTE while the Carrier 2*

will be using Cellular data of Carrier 1 on your iPhone.

At the Homescreen you will only see the active Cellular Network Status Bar Icon at the top right of the screen.

How to Setup Cellular Plan Manually

If you have already known all the details of the Cellular Number then take the following steps:

On the Add Cellular Plan Page: At the bottom of the page tap on the Enter Details Manually to open Enter Activation Code page.

On the Enter Activation Code Page:

- Enter SM-DP Address.
- Enter Activation Code
- Enter Confirmation Code by using the Keyboard below.

Note: It is necessary to enter a confirmation code if your carrier has issued one.

You can only use one SIM at a time on your iPhone because it has two SIM slots. As a result, before you can access the second SIM, you must switch from one SIM to another.

How to Use Dual Nano-SIM Cards on Your iPhone

- Set your iPhone down on a table with the screen facing up.
- Pick up the SIM Ejector, identify the hole on the SIM Tray that has been perforated by looking at the left side of the iPhone, and insert it there.
- Apply a little pressure to push the internal surface once the Ejector's head reaches it. The SIM tray will immediately nudge out.
- Pull the SIM Tray completely out using your two fingers.
- When the SIM Tray is turned upside down, a spring located at the left side end will prevent the nano-SIM Card from falling out. The initial nano-SIM Card's metal surface should be facing upward.
- To attach the card into the proper shape on the SIM Tray, push the spring inward with your SIM card.
- As illustrated in the illustration below, flip the tray over and turn the second nano-SIM Card's metal surface up.
- Place the SIM on the SIM Tray by turning it so that it fits in the proper shape. Make sure the SIM is balanced appropriately on the tray.
- By pushing it in, place the Tray within the SIM slot.
- Make sure the SIM Tray's surface is absolutely parallel (level) with the iPhone's side surface. To ensure the iPhone is perfectly level, move your finger around the tray's surface.

Note: If you secured your SIMs with Personal Identification Numbers (PIN), you should be able to enter each of the PINs correctly when prompted to enter them during settings, so you must be aware of the precise SIM you fixed to the front of the Tray and the other SIM you fixed to the underside of the Tray.

Chapter 6: - Customizing Your iPhone

Customize iPhone Home Screen

Your iPhone home screen should be set up in a way that makes sense to you, which means that you should be able to customize it to your heart's content. Customize your iPhone home screen with the options listed below.

Change the Background Image:

You can modify the home screen's background image however you like. You may use a picture of you, your family, or your spouse, or even the emblem of the company that most appeals to you. Go to Settings->Wallpaper. To access the wallpaper options, choose a New Wallpaper..

Use a Live Wallpaper or a Video Background:

Is there anything in particular you'd like to draw your attention to? As an alternative, consider using animated desktop backgrounds. It's a nice idea, despite certain limitations. Tap and choose Active or Live under *Settings > Wallpaper > Select a New Wallpaper*.

Get Organized: Make Folders for All Your Apps

Use *folders* to categorize your home screen apps according to how frequently you use them. Tap and hold a single app for a few seconds at a time until all of your apps start shaking. Once you've done that, drag and drop the two apps into a folder of your choosing.

Insert Pages into Apps:

Not all of your apps need to be on the same home screen. You can tap, hold, and drag applications or folders to create distinct "pages" for various users or application types on the right side of the screen.

Ringtones and Text Message Tones for the iPhone

It's not necessary for your iPhone's ringtones and message tones to sound exactly like everyone else's for them to be effective. You can tell who is calling or texting without ever looking at your phone because you have complete control over the sound.

To make a different ringtone the default, follow these steps:

Your iPhone comes with a lot of ringtones already installed. Change the standard ringtone for all incoming calls to the one you prefer to hear when you get a call. Choose "Noise" and "Ringtone" from the settings menu on your phone (or "Noise & Haptics" on some models).

Make your own ringtones:

Your trusty group can each have their own distinctive ringtone. Before you answer your phone when your partner calls, you'll hear a love song. If you want to change the ringtone for a specific contact, go to *Phone Settings > Contacts > touch on that contact > Edit > Ringtone.*

Incoming phone calls now have the option to be seen in their entirety. A boring incoming call screen is not essential. This concept enables you to observe the caller in full-screen mode. Click on *"Contacts"* in the phone's settings, then *"Edit"* under *"Add Picture."*

Message Tone Customization:

For calls, you can program a custom ringtone; for texts, you can program a video to play when you receive a message. Go to *Sounds (Noises & Haptics on some models) => Text Message Tone in the Configuration Settings.*

You have more options in addition to the built-in band and text tone on the iPhone. Apple offers pre-made ringtones for sale, and several third-party apps allow you to create your own.

Changing the Lock Screen on an iPhone

It's standard practice to personalize your iPhone's home screen and lock screen. As a result, you have total control over what is displayed on your phone when you wake it up.

Use the Lock Screen Wallpaper Customization option:

Like on the home screen, you may change the iPhone lock screen wallpaper by selecting a picture, computer animation, or video. The home screen should adhere to the same rules as those listed in the section above.

In order to create a more secure password,

The more difficult it is to gain access to your iPhone, the more complicated your *passcode* is. Depending on your iOS version, the default passcode length is four or six characters. You have the option to increase the length and strength of your passcode. Go to *Settings, Face ID (or Touch ID), and Passcode, Default Password*. Changing the password and following the on-screen instructions are recommended.

Get Suggestions from Siri:

Siri can use this information in a variety of ways to assist you in finding what you're looking for. Settings => Search & Siri => By setting specific settings to "on" or "green," Siri Recommendations gives you the option to ignore Siri's recommendations.

The Best iPhone Modifications:

These iPhone adjustments make it much simpler to notice what is happening, even if it isn't just a basic text message or on-screen item.

Focus on the Screen:

What if the text and icons on your screen are too small to be read with your eyes? Your iPhone's display size is automatically enlarged with Screen Focus. Go to *Settings => Screen & Brightness => View =>Zoomed => Collection to make use of this feature.*

Alter the font size:

You can increase the text size on your iPhone if the default size is uncomfortable for your eyes to read. You can make your text messages bigger by moving the slider to the On/green position and then moving the slider below.

Use the Dark Mode:

If the vibrant colors on the screen are too much for your eyes, you can select the "Dark Setting" option on the iPhone. *Configurations => provides access to the necessary Dark settings. Convenience > Screen Accommodations > Invert Colors.*

Customized Notifications on Your iPhone

You may be informed of new calls, messages, emails, and other pertinent information using the built-in notification system on your iPhone. On the other hand, the notifications sometimes cause problems. By using these suggestions, you can customize the notifications you get.

Decide on the notification method that's right for you:

From basic pop-ups to a combination of sound and text notifications, the iPhone offers a variety of notification options. Using *Settings->Notifications->tap the app you want to control*, select Alerts, *Banner Styles, Noises, and more.*

App-Based Notifications for a Group of People

Do you really need to see each notification taking up space on the screen if you receive many from a single app? A "stack" of alerts that occupy the same amount of space as your notice can be made. Go to *Settings->Notifications*, choose the app you'd like to manage, and then select *Notification Grouping*.

Notifications get a burst of light from Adobe:

As an alternative to receiving a notification, you can make the camera adobe. In most instances, it's a delicate but obvious choice. Make sure the LED Screen for Notifications is turned on under *Settings->General->Convenience->Hearing*.

Get a sneak peek at notifications using Face ID:

If you have Face ID on your iPhone, you may use it to keep your notifications secure. In notifications, you'll see a headline, but when you move through the screen and use Face ID to identify yourself, the notice expands to show more material. *Setting->Notifications->Show Previews->When unlocked, can be established.*

As an added bonus, the *"Reduce Alarm Volume and Keep Screen Shiny with Attention Awareness"* recommendation found at that link is also worth noting.

Other Options for Personalizing Your iPhone

The following are some additional techniques for customizing your iPhone.

Get Rid of Installed Applications:

Have you ever used any of the pre-installed apps on your iPhone? Most of them can be discarded if you don't like them. Simply perform the regular process of holding down the touch till they tremble, then tapping the x on the icon, to remove apps.

Change the look and feel of the Control Center.

There are a lot more options in the Control Centre than meets the eye. *Customize* the Control Center to obtain only the tools you need at your disposal. Go to *Settings->Control Center->Customize Settings* in order to change your preferences.

Install the keyboard of your choice:

Third-party keyboards, such as *Google search, emoticons, GIFs, and more*, may be installed on the iPhone, which already has an outstanding touchscreen keypad. Go to *Settings->General->Keyboard->Keyboards* and download a new keyboard from the App Store.

Take the time to get to know Siri:

You want Siri to use a man's voice when conversing with you? It's possible. Navigate to *"Settings" > "Siri" > "Siri Search." Voice type: Male*. Changing your accent is an option as well.

Changing the default search engine in your browser:

Is there another search engine besides Google that you would like to use? Set it as the default for similar searches in the browser. Make a new selection in *Settings->Browser->Search Engine.*

Make Your Own Shortcuts

If you're using an iPhone 11 or later version, you can create a range of custom gestures and shortcuts.

Become a Jailbreaker:

Jailbreaking your phone gives you full access to all of your phone's customization options, allowing you to bypass some of *Apple's restrictions*. If you're looking for more control over your phone, you may want to consider jailbreaking.

Chapter 7: - Setting Up Apple Pay

Methods for Apple Pay Identity Verification

The whole Apple Cash Family (Setting up for kids, view transactions, etc.).

Configure Apple Pay Cash (How to send money, payments, add money to the card, request payment, etc.).

Set Up Apple Pay

Comparing Apple Pay to an actual card is quicker and safer. You may use Apple Pay to make secure payments at shops, on public transit, in apps, and on websites that accept Apple Pay by using the Wallet app to store your cards. Utilize Messages to use Apple Pay to buy things from merchants who accept it while you're on the phone.

To start using Apple Pay, add your prepaid, debit, and credit cards to Wallet.

Add a Credit or Debit Card

Press the Add Card button in Wallet. You might be prompted to log in using your Apple ID.

Choose one of the following:

Previous cards: Cards linked to your Apple ID, cards used with Apple Pay on other devices, and cards removed from your wallet are all acceptable options. Press the Continue button to continue, and then input the CVV code for each card.

Debit or credit card: You can manually enter the card information or hold the iPhone so that your card is visible in the frame.

Transit card: Enter a location or card name, or scroll down to find transit cards in your region.

Set the Default Card and Reshuffle Your Cards

The first card you add to Wallet is your principal payment card. A card can be made the default by being moved to the top of the stack.

In Wallet, pick your default card, and hold down the card to bring it to the front of the stack.

To relocate another card, tap and hold it, then drag it to a new spot.

Utilize Apple Pay to Make Contactless Purchases

With the help of Apple Pay, you may use the credit, debit, and Apple Cash cards that are stored on your iPhone in the Wallet app to make secure, contactless payments at restaurants, stores, and other locations.

Pay Using Your Default Credit Card

Double-clicking the side button will make it active.

Use Face ID or your passcode to authenticate with your iPhone when your default card appears.

When you bring the iPhone up to a few millimeters from the contactless reader, the screen will display "Done" and a checkmark.

Pay Using a Different Card Than the One You Normally Use

When your default card appears, press it to access it, then choose a different one.

To authenticate, use Face ID, Touch ID, or your passcode.

When you move the iPhone up close to the contactless reader—a few millimeters—the screen will display "Done" and a checkmark.

Use Apple Cash

Your Apple Cash card in the Wallet app is replenished when you receive money over Messages. Using Apple Cash is as simple as using it anywhere you'd use Apple Pay. Additionally, you have the option of transferring your remaining Apple Cash to a bank account.

Set Up Apple Cash

Do any of the following:

- Go over to Settings.
- Click Wallet and Apple Pay, then turn on Apple Cash.

In Messages, send or receive a payment.

Use Apple Cash

You can use Apple Cash everywhere you use Apple Pay:

You may conveniently and quickly send and receive money using Apple Pay.

Apple Pay enables contactless transactions.

Use Apple Pay to make payments in applications or online.

Manage Your Apple Cash

In Wallet, tap the Apple Cash card.

View your current transactions, or scroll down to see all your transactions sorted by year.

Tap the More icon, then perform any of the following:
- Wallet money can be added using a debit card.
- Send funds to your bank.
- Update the information on your bank account.
- Ask for a statement.

Select whether to take all payments manually or automatically. You have seven days to manually accept the payment before it's sent back to the sender.

Take a look at your suggested PIN. There is no requirement for a PIN because every Apple Cash transaction is validated by Face ID, Touch ID, or a strong password. A four-digit number may still be required to complete the transaction on some terminals, though.

For account services, confirm your identity, and increase your transactional limitations.

Contact Apple Support.

Use Apple Card

To aid you in leading a more financially responsible lifestyle, Apple created the Apple Card credit card. Using the Wallet app on your iPhone, you can quickly sign up for an Apple Card and start using it with Apple Pay in stores, applications, and online throughout the world. Apple Card shows your recent transactions and balance in your Wallet in a straightforward, real-time manner, and you can contact Apple Card support at any time by sending a text message from inside Messages.

Purchase an Apple Card

Select Apply for Apple Card from the Wallet's Add Card menu.

Enter your information and confirm your agreement to the terms and conditions to submit your application.

Before accepting or rejecting your Apple Card offer, carefully read the terms and conditions, including the credit limit and annual percentage rate.

You are permitted to take the following steps if you accept the terms:

The default card for Apple Pay transactions need to be Apple Card.

To use in places that do not accept Apple Pay, get a real Apple Card.

Utilize the Apple Card

All Apple Pay establishments are compatible with Apple Card: To make contactless purchases, use Apple Pay.

You can make payments using Apple Pay both online and in applications.

In addition, you can use Apple Card in places that do not accept Apple Pay:

When making a phone call, utilizing an app, or browsing the web: To view the card number, expiration date, and password, tap the Card Details icon. Utilize these details to place an order.

In shops, restaurants, and other venues, use the actual card.

Pay Your Bills

Select the Payment option. Alternatively, hit the 'More' icon and then select one of the following:

Pay My Bill or Pay Different Amount: After entering the payment details (such as the account and date) and choosing Pay My Bill or Pay Different Amount, you can confirm with Face ID, Touch ID, or your passcode.

Pay a Bill: Drag the checkmark to alter the payment amount or tap Show Keypad to enter the amount, then select Pay Now or Pay Later, look over the payment details (such as the payment account), and then confirm using Face ID, Touch ID, or your password.

Chapter 8: - Apps

In this section, we will cover tips on using and operating apps on your iPhone 14.

Open Apps on iPhone

There are two ways to open an app: from the App Library or from the home screen pages.

- To access the Home Screen, swipe up from the bottom of your screen.

- To view the apps on the various home screen pages, swipe laterally.

-

- To access the App Library from the Home screen, swipe left all the way to the bottom of the page. Your apps are categorised in the App Library.

- To launch an app, click its icon.

- To get back to the App Library, swipe up from the bottom of your screen.

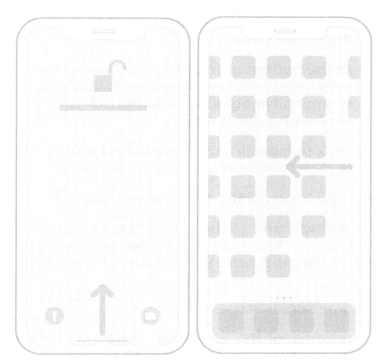

Find and Open Apps in the App Library

- You may access the App Library by swiping left from the Home screen all the way to the bottom of the page.

- To open an app, select it by clicking on it after typing its name into the search box at the top of your screen. You can also look for an app by using the alphabetical list.

- To view every app in a category, click on it to open it.

Show and Hide Home Screen Pages

You might not want to have as many apps on your home screen because they are all neatly arranged in the app library. You can access the App Library more quickly and easily if you hide some of the pages on your home screen. To hide

- When you notice the apps jiggling, press and hold the Home Screen.

- At the bottom of your screen, click the dots. The pages of your Home screen will appear as thumbnail images at the bottom of your screen, each with a checkmark.

- Click to include a checkmark and display the page, or click to remove it and hide the page.

- then to complete, click Done twice.

Note that hiding some of the Home Screen pages would mean that newly downloaded apps would automatically be added to the App Library only rather than the Home Screen.

Reorder Home Screen Pages

Change the order of the pages on your home screen so that it works best for you. For instance, you could put all of your favorite apps on one page and put that page at the top of the list.

- The apps will start to jiggle when you press and hold the Home Screen.

- Go to the end of your screen and click the dots. The pages of your Home screen will appear at the bottom of your screen as thumbnail images, each with a checkmark.

- You can drag a page to a new spot by pressing and holding the page you want to move.

- Then click Done two times to finish.

The App Library

The App Library's apps are divided into a number of categories, such as Entertainment, Creativity, and Social. These categories are produced by iPhone 14 based on how you use the apps on your phone. While moving an app from the App Library to the Home Screen pages is possible, moving an app from one category to another is not possible. To make it simple for you to access them, your frequently used apps will appear near the top of the screen and on the top portion of their category.

Change the Storage Location for New Apps

Apps that have just been downloaded can be put to the Home Screen and App Library, or just the App Library.

- In iPhone's Setting app, click Home Screen.
- Then choose whichever option you desire: either the Home Screen and App Library, or only the App Library.
- Turn on Show in App Library to enable app notification badges to show on the apps in the App Library.

Add Apps from the App Library to the Home Screen

An app may be in the App Library without being on the Home Screen page. To add such apps to the home screen page,

- Find the app in the App Library, press and hold the app and then select Add to Home Screen.

The App Switcher

All open apps are saved by the app switcher, which also gives you the option to switch between them. for a list of your open apps,

- Swipe upward from the bottom of the screen, pausing halfway up the middle.

- Move right and left until you find the desired app, click it to launch it, and you will see all open apps.

Switch Between Open Apps

Here is a fast method for navigating between open apps,

- While using one app, swipe left or right along the bottom edge of your screen to open other apps until you get the one you want.

Force Close Apps

You may quit an app that isn't responding and then reopen it to fix whatever the problem may be.

- Swipe upward on the app after opening the App Switcher and moving to the right until you reach it.
- then look for and launch the app using the App Library or Home screen pages.

Multitask with Picture in Picture

This is one great multitasking feature that allows you to perform two tasks at the same time, like operating an app while watching a video or doing FaceTime.

- Play the movie and then use your mouse to click the corner of the screen to minimize the video.
- To bring the video back to full screen, tap.
- To conceal the window, drag the video to the right or left side of your screen.
- While using the other app, move the video from one corner to another.
- To increase or decrease the size of the video window, pinch it open or closed.

Chapter 9: - Camera Basics

Learn how to use your phone's camera to snap images. Choose from many camera modes, including Cinematic, Photo, Portrait, and Video.

Open the Camera

To enter the Camera application, swipe to the left on your phone's lock screen, or click the Camera icon on your Home screen.

Note: For your protection, a green dot would appear in the upper right part of your screen when using the Camera.

Switch Between Camera Modes

Photo is the mode you see when you enter the Camera application. Swipe to the right or left to pick any of the following modes:

- Video: Record a video.
- Pano: Take a panoramic landscape.
- Timelapse: Create a timelapse video of motion over a period.
- Cinematic: Use depth of field effects on your videos.
- Slo-mo: Record a slow-motion video.
- Square: Take a picture with a square ratio.
- Portrait: Use depth of field effects on your pictures.

Zoom in and Out

You can zoom in in 2 ways.

- Enter the Camera application and pinch open or close to zoom.
- Toogle between 0.5x, 1x, 2x, 2.5x, & 3x to zoom. Hold down the zoom control and move the slider to the left or right for a more accurate one.

Take a Live Picture

A Live picture takes everything that happens before and after taking your picture, as well as the sound.

- Activate the camera's photo mode.
- Ensure that the Live Picture is turned on. When turned on, your camera's upper portion will display the Live Photo icon. To activate or disable Live Photo, touch the Live Photo button.
- A Live Picture can be taken by pressing the Shutter button.
- The Live Photo can be viewed by clicking on the thumbnail below the screen and then holding down the screen button to play it.

Take a Photo with Burst Mode

Snapping moving objects is possible while using the burst mode. You have a large selection of images to choose from because burst mode takes numerous high-speed photographs. The front and back cameras both supports burst photo capture.

- For snappy photos, swipe the shutter to the left.
- You can see how many photos you take on the counter.
- To halt, lift your finger.
- To select the images you want to save, tap Select after on the Burst thumbnail.
- The suggested images to keep are indicated by gray dots beneath the thumbnail.
- To save multiple images as one, select Done after clicking the circle in the bottom right corner of each image.
- To erase every picture from the Burst mode, click the thumbnail and then tap the Delete button.

Take Apple ProRAW

Apple ProRAW joins the info of a standard RAW format with your Phone's image processing to provide extra creative control when you adjust color, white balance, and exposure.

- Go to Settings > Camera > Formats and turn on Apple ProRAW to configure Apple ProRAW.
- Use Apple ProRAW to take a photo.
- Touch the ProRAW button in the Camera application to start it up.
- Take a photograph
- To enable or disable ProRAW while shooting, alternate between the Raw On key and the Raw Off key.

Record a Video

- Opt for Video mode.

- To begin recording, tap the record button or use any of the volume controls. You can perform any of the following when you record:
- To capture a picture, press the white shutter.
- To zoom in, pinch your screen.
- To stop recording, press the Record button..

Record a Slow-Motion Video

When you record a video in the Slo-mo mode, you simply record the video as normal, and when you replay it, the slow-motion effect is visible. You can also modify your video so that the slow-motion effect begins and ends at predetermined times.

- Select the Slo-mo mode.
- Touch the Camera Selector button to record the Slo-mo with the front camera.
- To begin recording, tap the Recording button.
- As you record, you can push the "Shutter" button to capture a photo.
- To stop recording, tap the Recording button.
- Touch the video thumbnail and select Edit to have one part of the video play slowly while the remainder plays at a consistent speed. The portion you want to play can be selected by slowly dragging the vertical lines underneath the frame viewer.

Use Live Text with Your iPhone Camera

Your iPhone camera can copy and share text, open webpages, and make phone calls from the text inside the camera.

- Place the iPhone so that the text is visible on your screen and open the camera.

- Click the Live Text button when the visible text is surrounded by a yellow frame.
- To highlight text, swipe or use the grab point. Then carry out any of the following:
- Text can be copied and then pasted in another location.
- Choose All: Choose the entire frame's text.
- Look-up: searching for text online.
- Translate.
- Share:
- Text can be shared using Mail, Messages, AirDrop, etc.
- To access a website, dial a number, or start an email, On your screen, click the website, phone number, or email address.
- For the camera to return, tap the Live Text On button.

Scan a QR Code on Your Phone Camera

You can scan Quick Response (QR) codes with your camera to find links to websites, applications, tickets, coupons, and other content. The camera would recognize and pick out the QR code.

- Open the Camera app, then position your phone so that the barcode appears on the screen.
- To visit the relevant website or application, click the notification that will appear on your display.

How to Shoot Cinematic Video

With the iPhone 14 and subsequent models, Apple introduced Cinematic Mode, a novel method of shooting video that lets you easily adjust focus and track objects both while filming and after.

What does Cinematic Mode accomplish?

While a movie is being recorded or subsequently, Cinematic Mode enables continuous depth of field correction. In filmmaking, the terms "racking focus" and "pulling focus" refer to changing the emphasis on a certain subject or object to divert the audience's attention.

In this mode, the iPhone 14 can record videos in Dolby Vision HDR up to 1080p at 30 frames per second. The 24 frames per second (24p) frame rate used by the vast majority of theatrical films means that the frame rate constraint shouldn't be a serious issue, but better video would have been nice.

After shooting your movie, you may use keyframes to apply focus pulls at specific intervals, basically allowing you to focus on any object in the frame that has been in focus for the duration of the movie.

Apple claims that it has improved its focusing algorithms so that they now automatically identify and monitor any objects you might also want to focus on. By touching a person or object in the frame, the functionality can be turned on. When you tap the screen once more, the camera will track that object and show the message "AF Tracking Lock."

Apple claims that when users do things like gaze away from the camera, the iPhone 14 anticipates them approaching the screen and instantly switches focus away from them.

Pano Pictures

Panorama Mode allows you to capture a panorama image from your iPhone or iPad camera by stitching together a series of photos taken from different sides of the camera. There are 4 different panorama photography modes:

- Pano
- Horizontal panoramic
- Vertical panoramic
- 360 panoramic

To remove an effect from a photo, do the following:

- Open the photo. The photo opens, and you've returned to the Camera screen.
- Click an image you would like to get rid of its filter.

- Press the Edit option, and then tap Revert →Revert to Original.

The photo no longer has the effect applied.

Chapter 10: - Siri

Ask Siri

Utilize Siri to complete tasks swiftly. You may use Siri to set an alarm, obtain directions, check the weather, or translate a sentence. See Learn more about the capabilities of the iPhone's Siri by using it.

Although voice input is processed by the iPhone, Apple receives the transcripts of your requests so that Siri can be enhanced. This information won't be kept around for very long, unlike your Apple ID. Apple could potentially use the speech recordings you provide to improve their products. More details are available at Improve Siri and Dictation & Privacy.

Your iPhone must be online to complete some requests. Possibly included are cellular expenses.

Set up Siri

To activate Siri on your iPhone, go to Settings > Siri & Search and perform one or more of the following:

- If you want to activate Siri by speaking commands to it, follow these steps: Turn on the "Hey Siri" listening mode.
- For an iPhone with Face ID, you can use the Side Button to summon Siri, or press Home to summon Siri using a button (on an iPhone with a Home button).

Tell Siri about yourself

For a more customized experience, you may offer Siri with information such as your home and work addresses, and your connections, so you can say things like "Give me driving instructions home" and "FaceTime mom".

Tell Siri who you are

1. Open the Contacts app, then enter your personal details.
2. Then tap on your name under My Information in Settings > Siri & Search > My Information.

Tell Siri how to say your name

1. Tap your contact card to bring up the Contacts app.
2. Add a new name field by tapping "add field," then put your name in the field.

Have Siri announce calls and notifications

Siri may announce incoming phone calls and alerts from programs like Messages while using compatible headphones or CarPlay. You don't need to say "Hey Siri" in order to reply or respond vocally.

Announce Calls and Announce Notifications can be used by third-party applications as well.

Have Siri announce calls

Approve or reject phone calls and FaceTime chats by speaking to Siri, us-ing Announce Calls.

1. Choose an option from the drop-down menu under Settings > Siri & Search > Call Announcements.
2. Incoming calls are answered by Siri by asking whether you'd want to take the call. To accept or reject the call, just say "yes" or "no".

Have Siri announce notifications

Siri may automatically announce notifications for reminders and messages. Siri automatically triggers app notifications for programs that use time-sensitive alerts, but you may easily change the settings. For additional information on time-sensitive notifications, see Set up Focus on iPhone..

1. Make sure Announce Notifications is enabled in Settings > Siri & Search.
2. Turn on Announce Alerts for the app you want Siri to announce notifications from.

However, some applications provide you the option to choose between viewing all alerts or just those that are time-sensitive.

In programs like Messages, Siri repeats what you said and then requests confirmation before responding. Reply Without Confirmation should be turned on to deliver responses instantly.

Add Siri shortcuts

For typical tasks, several apps offer shortcuts. Siri, a voice-activated assistant, makes using these shortcuts straightforward. Ask Siri, "Where am I heading next?" and you'll see your upcoming itinerary right away if you have a travel app.

When a suggested shortcut appears, select Add to Siri and then adhere to the onscreen instructions to record a word or phrase of your choice to use the shortcut.

Additionally, you can utilize the Shortcuts app to manage, re-record, and delete already-existing Siri shortcuts in addition to creating new ones.

Use a shortcut

Simply enable Siri and then speak your sentence to use the shortcut.

Additionally, Siri offers shortcuts for your preferred programs depending on how frequently you use them and how frequently you begin searches.

How to use Siri in your car

With CarPlay or Siri Eyes Free, you can use Siri while you're driving to make calls, send texts, play music, get directions, and use other iPhone services. You can use your iPhone's built-in display to complete tasks while driving if your car supports CarPlay (available in select models). You can use your voice to handle CarPlay because Siri is utilized to manage it. See How to Use Siri for CarPlay Control.

You can use Siri Eyes Free in your car to control your iPhone's features using just your voice, without having to look at or touch it. Your iPhone and vehicle can be connected via Bluetooth (refer to the user guide that came with your car if you need to). In order to speak to

Siri, press and hold the voice command button on your steering wheel until you hear Siri's tone.

How to change Siri settings

You may alter Siri's voice and prevent access to it while your phone is locked.

Change when Siri responds

Enter Settings > Siri & Search, then select one of the following options:
- If Siri is turned off, it won't respond to the "Hey Siri" command. Switch off Keep an eye out for the command "Hey Siri."
- If you do this, Siri will no longer respond when you hit the side or Home button. On an iPhone with Face ID, press the Side button to turn off Siri. On an iPhone without Face ID, hit the Home button (on an iPhone with a Home button).
- Siri cannot be used when the phone is locked. You can disable Allow Siri When Locked.
- You can modify the language in which Siri responds: To switch to a different language, click on it.

Change how Siri responds

Enter Settings > Siri & Search, then select one of the following options:
- Replace Siri's default voice with: (not available in all languages) Siri Voice may be used to choose a different voice or variation.
- Change when Siri responds to your spoken

commands: Siri Responses may be accessed by tapping Siri Responses, then selecting the Spoken Responses option.
- Keep an eye out for Siri's reaction at all times: Turn on Always Show Siri Captions by tapping Siri Responses and then selecting Always Show Siri Captions.
- Check out your request here: Take a look at Siri Responses, and then choose Always Show Speech.

Change Siri settings for a specific app

All of the Siri Suggestions settings and Siri Shortcuts settings for any app may be changed.

- Scroll down in Settings > Siri & Search to find the app you want to use.
- Activate or deactivate a setting.

Retrain Siri with your voice

Turn off "Hey Siri" listening in Settings > Siri & Search, then turn it back on in Listen for "Hey Siri."

Chapter 11: - First-Hand Tips and Tricks

We'll look at the first 13 things you should do with your brand-new iPhone 14 or another model in the iPhone 14 series; these techniques, tips, and features will ensure that you have a solid setup and that you get the most out of ownership. So let's jump in and begin straight away!

Battery

The first thing you should do is head to settings, scroll down till you see battery, and then click on battery. Then, the first thing you should do is tap on battery health.

Make sure the maximum capacity is 100 percent when you access the battery status; if it's 99 percent when you first acquired the phone, that's a concern even though it's not a poor number because it suggests the batteries has been utilized in some way.

The minimum capacity on a brand-new iPhone should be 100%, and below that, you'll notice top performance potential. If you get a used phone and the warranty is 90 percent, it is not a big concern, but if it is a brand-new phone, you should return it and purchase a new one. Check to see if optimal battery charging is enabled when inside the battery. When you do that, the charging speed is going to be a little slower, but it's also going to increase the battery life of your phone over the course of your ownership. If you disable it, however, what will happen is that the phone is going to charge a little faster; once it reaches 80%, it keeps charging faster, but if you have the optimized battery charging enabled, once the phone reaches 80%, it starts to charge slowly and, believe it or not, this gives you more battery life.

Auto Brightness

Turning off the auto brightness option on your phone

should be the next thing you do in the main settings. If you slightly lower the main setting, select Display and Brightness.

As you can see from the image above, we have a brightness adjuster, but for some reason the toggle to turn on and off automatic brightness is concealed. Going back to options, scrolling down a bit, selecting accessibility, selecting display and text size, and making sure auto brightness is off at the bottom should be your next steps.

When it's turned on, the phone controls the auto brightness itself, which may be convenient for some users but is more frequently used by others who prefer manual control. Therefore, as it is set to be on by default, switch off auto brightness. While you are there, you should also rapidly adjust the bold text. So, if you slightly scroll up, you can see that the bold text is activated.

The sentences will be a little bit thinner if you disable it, making it easier and more pleasant for you to read.

Raise to Wake

The next parameter you need adjust is called "raise to wake feature" and it can be found under display and brightness as well. If you want the option, make sure that it is turned on. By raising your phone to your face, you can wake it up using the raise to wake feature.

You should turn it off if you want to conserve a little bit of battery life because every time you raise the weight, the screen comes on and drains your power.

Auto Lock

Another thing that many people overlook is the auto lock feature. Ensure that auto lock is set to either 30 or 1 minute, which will cause the phone to switch off after that amount of time. If you have the auto lock set to five minutes or four minutes, the screen will stay on and continue to drain your battery. When I say it shuts off, I mean the screen goes dark.

Sounds and Haptics

One more battery related feature: if you go back to your setting and tap on sounds and haptics

What you should do is to disable vibrate on ring. The reason you should disable this is because the vibration engine actually wastes some battery, when the phone is ringing, you can hear the call so you don't need it to be vibrating, that way you're going to save some battery life but you can keep the vibrate on silent enabled, so that when the phone is silent and somebody calls you or sends you a message then the phone vibrates which makes sense.

Extract Text Feature

The extract text feature is the next topic I want to cover. The image below features a photo with text written on it.

There are a few methods for extracting text, and I'll demonstrate them all for you. Press and hold anywhere on a photo when using Safari to bring up a menu of options; select show text from the list.

When you click on "Show Text," a new window will open and the image's text will be selected.

Now, all you have to do to copy that text is press and hold on it while choosing the portion you want to actually copy; from there, you may copy and paste it wherever you choose. So that's one method of extracting information; another is to simply press and hold on the text. From there, you may select the text you want and copy, resize, look it up, or translate it.

The last method of text extraction is through photo downloads. For example, imagine you are in a school and you take a picture of a whiteboard with some text on it. All you have to do is press and hold once you have the picture to begin selecting the text on it.

Security

Let's talk about your phone's security; it's crucial to secure the phone properly. Go to your settings, scroll down a bit, press on facial identification and passcode, and then enter your password.

Ensure that each and every option is selected in the image below.

Your face ID can be used to unlock your phone, make iTunes purchases, use Apple Pay, and, if necessary, auto-fill your password.

It's vital to note that the "need attention for face id" and "attention aware features" options must both be activated. Scroll to the bottom of the page to find them. Let's imagine someone steals your phone while you are sleeping. If the "need attention for face id" feature is off, your girlfriend may take your phone to see who you have been texting that day. In such case, she can just hold the phone up to your face to unlock it. If this option is deactivated, your eyes may be closed or you could be looking somewhere else, which would allow someone to unlock your phone using your face if you were having a nap. If it is enabled, your eyes must be open, and you must be looking at the sensor.

Make sure to set up an other look as well. For example, you can wear a mask and set an alternate appearance to make it simpler to unlock your phone when you're out and about. Other options include wearing a wig, sunglasses, or a mask.

At the bottom, which is quite significant, is the allow access while locked option.

All of your phone's enabled capabilities are accessible to anyone when it is locked. Therefore, be cautious to turn off any delicate features and just enable those you feel comfortable using. Ensure that the phone is completely safe!

Restrictions

If you frequently lend out your phone to others, you should set up limitations to prevent damage or unintended purchases. For example, you might wish to prevent access to particular components of the phone. What you need to do is go your settings and select screen time.

When you access this area for the first time, it could prompt you to set up a passcode; when you do, make sure it's separate from the passcode for your lock screen. Scroll down and tap on content and privacy controls. Assuming you have the passcode set up, you can simply tap on the authorized apps and enter the passcode to block particular applications.

After that, you can restrict access to all of the programs listed.

The apps on your home screen will no longer be accessible when you disable access to any of them.

The purchases made through the iTunes and app stores are a crucial factor in this regard. Do you initially want to allow others to install programs when you tap on it? If not, choose no and check the box next to Do you want others to be able to delete your applications? In that case, choose no.

Since that option has been removed, users may no longer press and hold to delete programs.

In-app purchases are the huge thing I like, so make sure you forbid them. If you permit this and give someone your phone while they're playing a game, they may unintentionally make in-app purchases and spend your money.

In fact, you might just want to turn off in-app purchases to save money because we sometimes buy things in apps way too much. These are the app limits that you should check to see if they are properly configured, or at the very least, configure before giving someone your phone.

You don't want to be on social media all the time, so make sure you set certain limits for specific social media applications by selecting app limits. That's one more thing I want to demonstrate to you here that is extremely cool. Therefore, press social before selecting next.

The restriction will be added after you choose how many hours per day you wish to spend on social media and click the "add" button. As a result, the phone will restrict your access to social media after the allotted time has passed. You can still bypass this restriction, but at least it will serve as a reminder.

Freeing Up Space

Go to your settings, click general from the menu, then choose iPhone storage. Make sure "offload unused applications" is enabled once you're in your iPhone's storage.

This basically offloads unused applications, so if you haven't used an app for days, weeks, or months after installing it, the phone will eventually learn to do so and will eventually offload it, giving you more space for your images, movies, or other apps that you are currently using.

Control Center

Your control center is the next item you'll frequently use, so make sure you give it the appropriate customization.

The first three components on the left cannot be altered, but the remaining components are entirely editable. For more information, press and hold the first three components on the left that cannot be customized.
If you wish to add a control to your control center, simply click the plus button after selecting Control Center from the Settings menu.

Make sure to add the toggles you desire, and if there is one toggle in particular that you know you will use frequently, take it and move it to the top. Simply press the side and place it at the top to grab it.

Camera

Since I'm expecting you'll be using your camera frequently, when you run the camera application, you can change between modes by sliding left or right, but you are unable to reach the camera settings. The settings and setups for your camera will therefore be available if you go to your settings, scroll down to the camera, and tap on it.

Make some simple adjustments, touch on record video, and ensure that the video is set to 1080 HD at 60 frames per second because it is the industry standard, it takes up less space, and it is very smooth. If you want to do 4K, you may select it from the options.

HDR video can be enabled or disabled; most users choose to disable it. After that, they make sure the grid is enabled so that when they open their camera app, the grid lines are present for better composition.

One more thing: in your camera app, under video, you can adjust the resolution from 4K to 1080p and the frame rate from the top.

Privacy

Privacy is crucial, so make sure you are aware of what is occurring with your phone and the programs you download. Always access your options, scroll down, and select privacy.

The key elements of your phone are shown in the graphic above, and you can see what the applications have access to by looking at it. Consider your camera, microphone, and Bluetooth. If you click on the camera, for instance, a list of all the programs that have visited your camera will appear. You may click on any of them to make changes or to read descriptions.

Make sure you properly control your privacy by being wary of the apps and how they collect data about you!

Safety, Use and Care information for the iPhone

- On a level, sturdy surface, the iPhone setup should be performed.
- Make sure the iPhone is kept away from any liquids at all times (dironks, drinking water, shower, bathtubs). The device could sustain damage if water gets on it.
- Pull on the plug to disconnect the electricity, not the cord.
- The base of the iPhone typically warms up when in use. Therefore, avoid leaving items on your laps for an extended amount of time since this could result in a burn.
- The ports are designed to accommodate connectors with ease. A connector should not be forced into a port. A connector was presumably not designed to fit into a port if it does not fit into it easily.
- Turn off the iPhone and unplug the power adapter before cleaning the phone's exterior or its screen. Please avoid using aerosols, sprays, or abrasive materials and instead use a soft, moist (with water) cloth or cotton.
- The device's battery is fitted into the back. Under no circumstances should you try to remove or replace the battery yourself. An Apple Engineer or Authorized Apple Products

Service provider should replace the battery. Additionally, shield the device from heat sources (above 1000 c) to prevent the battery from igniting. It is advised to use operating temperatures between 100 and 350 °C and storage temperatures between -250 and 450 °C.

- Never try to open, battery-change, or otherwise fix an iPhone. For any servicing or repairs that require dismantling the device, please contact an Apple repair center or Service Centers Authorized by Apple. The warranty may be void if the gadget is opened or serviced without authorization.

- When using the device to play a media file, avoid exposure to loud noises. Hearing loss may result from loud music.

- Keep iPhone accessories out of the reach of young children.

Chapter 12: -Finding Lost Devices

on iPhone 14

When your smartphone is dead, can you still utilize Find My iPhone? You won't be able to see the position in real time, though, so yes. A misplaced iPhone can be located using Bluetooth even if the power is dead or Wi-Fi and cellular connectivity are unavailable. With the help of a number of tools and techniques, we'll show you how to find your iPhone's current position. Start the process now!

You can use a Mac computer, which is what you already own if you have an iPad, to find your lost iPhone. Just make sure your Internet connection is up and running. The Find My iPhone app must first be launched. From the list of devices, select your iPhone by tapping it after selecting Devices in the lower right corner of the screen.

A smartphone with an empty screen will appear on the map and sidebar if your phone is off.

A vivid (animated) Start screen can be seen when your iPhone is turned on.

Press Directions once more to get driving or walking directions to your iPhone's precise location.

Use Play Sound to help you locate your iPhone when you're nearby if it's off but not empty.

If your iPhone isn't in use, you can activate Notify When Found to send notifications and a location report to your iPad the moment it comes back on.

Reachability Mode on iPhone 14

Because iPhone screens have gotten bigger, using one hand to operate one of them occasionally may be challenging. If you're having trouble accessing items near the top of your iPhone's display, consider giving your overworked fingers a break and starting to use Apple's Reachability feature to pull the upper half of the screen within reach. You must enable Reachability in the Settings app in order to use this feature. Here's how you can make Reachability work for you.

1. Open up the Settings app.

2. Tap Accessibility> Touch (or just tap it if you know what you are looking for).

3.   Scroll down until you see the option called Reachability, then turn that toggle switch on.

4.   When you've finished, return to the main settings menu and scroll down to where you turned off Reachability. Now there should be an icon labeled More Gestures. Turn on the option under More Gestures so you can access the new gesture controls. You may also want to adjust some other accessibility options like VoiceOver or Switch Control depending on what kind of phone you have:

5.   After enabling these features, go ahead and try out all the gestures listed below. I'm sure you'll find them useful!
- To access the home screen, tap from the left edge.
- To advance to the next page, swipe right.
- To return a page, swipe left.
- To exit an app, double-swipe downward.
- To close an app, swipe down three times.
- Drag up to place applications in directories.
- To move an object between two pages, hold down.
- To add something to your favorites, long press.
- To remove something from your favorites, tap briefly.
- To zoom in or out, pinch.
- To move while zooming, flick.
- Lastly, pinch images to rotate them 90 degrees clockwise.

You may always utilize Siri to handle the most of the job if you'd prefer not to fiddle with any of these settings. To display your list of friends' names, just say something like, "Hey, Siri, show me my contacts." Then say "Call Lynn," "Call Kim," or even "Call Sasha" to indicate to her who you want to call. Or you may ask Siri to send an SMS or set an alarm.

message, play music, or anything else she can handle. It's pretty handy when you don't feel like messing with the settings yourself.

Recording Dolby Vision Videos on iPhone 14

This is a crucial feature for any photographer who uses their phone as a light meter or shoots in the outdoors because it enables you to capture more detail without having to worry about clipping highlights. Additionally, Apple Images will be able to display that high-dynamic range film far better than in earlier versions of iOS, which is wonderful if you want to edit your photos with that program.

Other than that, there are a few features that I believe will make filming videos much simpler. The first is Night mode, which lowers screen fatigue by reducing brightness when nighttime video is being recorded. If you wish to photograph a wedding outside but don't have enough time during the day, you should still be able to use Night Mode to avoid becoming sick because it also works really well indoors.

Turning On Dolby Vision HDR

Open the Settings app, then select the "Camera" icon to the left of the Photos button to toggle between the standard and High Dynamic Range (HDR) settings.

After you've done that, select Record Video from the menu. You should notice a difference right away if you switch on HDR Video mode. The quality of your photograph will be sharper and brighter than before. However, your videos won't appear as bright if you choose not to use high dynamic range.

You must manually adjust the power level of an external flash if you're using one. If not, there shouldn't be anything else you need to change.

Final Words

In order to be here, you must have finished reading this book and be extremely agitated. I'm sure you're wondering, "Where do I even begin?"

First, give each change, option, or piece of advise a try that interests you. After using it for a while, choose whether you want to keep it as a permanent addition to your iPhone or get rid of it.

The iPhone 14 series is Apple's strongest lineup of iPhones to yet, and it is technologically advanced enough to compete with any other smartphone on the market. As you grow more acclimated to your smartphone's numerous features and functionalities, you'll love it even more. I'm grateful you purchased the book and took the time to read it!

Made in the USA
Monee, IL
05 November 2022

17026013R00056